The Amazon and the Sahara

Using Double Line Graphs and Double Bar Graphs

Orli Zuravicky

PowerMath™

The Rosen Publishing Group's
PowerKids Press™
New York

Published in 2004 by The Rosen Publishing Group, Inc.
29 East 21st Street, New York, NY 10010

Book Design: Michael J. Flynn

Photo Credits: Cover (Sahara), pp. 11, 15, 19 (gazelle, camel), 23 (palm trees) © Royalty-Free/Corbis; cover (Amazon) © Theo Allofs/The Image Bank; pp. 5 (Amazon), 13 © Wolfgang Kaehler; p. 5 (Sahara) © Christine Osbourne/Corbis; p. 7 (San Diego) © Bill Ross/Corbis; p. 7 (Seattle) © Jim Corwin/Index Stock; p. 9 © Steve Vidler/SuperStock; p. 17 (jaguar) © Tom Brakefield; p. 17 (parrots) © Kevin Schafer/Corbis; p. 17 (otter) © Nicole Dublaix/Corbis; p. 19 (addax) © Steve Kaufman/Corbis; p. 21 (cinchona tree) © Hulton-Deutsch Collection/Corbis; p. 21 (rubber tree) © Sheldan Collins/Corbis; p. 21 (cacao beans) © Bob Krist/Corbis; p. 23 (background) © Fred Carol/Witness/Corbis Sygma; p. 25 (Amazon) © Collart Herve/Corbis Sygma; p. 25 (Yanomami) © Corbis Sygma; p. 27 (people) © Patrick Ward/Corbis; p. 27 (oasis) © Robert Holmes/Corbis; p. 29 © Jose Fuste Raga/Corbis; borders throughout © Digital Vision.

Library of Congress Cataloging-in-Publication Data

Zuravicky, Orli.
The Amazon and the Sahara : using double line graphs and double bar graphs / Orli Zuravicky.
v. cm. — (PowerMath)
Includes index.
Contents: Graphing the world — A quick look at the Amazon and the Sahara — A difference in climate — Worldly wildlife — Magical plants — The people of the Amazon and the Sahara — Changing ecosystems.
ISBN 0-8239-8981-X (lib. bdg.)
ISBN 0-8239-8868-6 (pbk.)
6-pack ISBN: 0-8239-7376-X
1. Graphic methods—Juvenile literature. 2. Amazon River Region—Juvenile literature. 3. Sahara—Juvenile literature. [1. Graphic methods. 2. Amazon River Region. 3. Sahara.] I. Title. II. Series.
QA90.Z87 2004
511'.5—dc21

2003002683

Manufactured in the United States of America

Glossary

Graphing the World

Everything around you is a part of your community, or **ecosystem**. An ecosystem is a community of living plants and animals. An ecosystem also includes things like the soil, water, air, and weather of a particular region. In an ecosystem, a change in 1 part of the system can cause changes in other parts of the system. In order to understand these changes, **mathematicians** have designed many ways to record and study the **data** that is collected from observing all of these relationships.

One way to see and organize data is on a graph. Graphs are used to illustrate different types of data and make the data easier to understand. This book will teach you how to use 2 different types of graphs to discover the amazing relationships that exist inside 2 very different ecosystems—the Amazon rain forest and the Sahara desert.

The Amazon rain forest is located in South America. The Sahara desert is in northern Africa.

Amazon rain forest

Sahara desert

A double bar graph shows 2 bars side by side that compare the quantities of 2 different things. Each bar on the graph represents a certain value. The bars of a bar graph are often in different colors to help readers tell them apart more easily. The things being compared are usually labeled on the bottom of the graph. The **units** of measurement—such as pounds or miles—are labeled along the side of the graph. The double bar graph on page 7 shows the average annual rainfall for Seattle, Washington, and San Diego, California.

A double line graph shows how things change over a period of time. The points on the double line graph on page 7 show the average monthly rainfall in the cities of Seattle and San Diego. The graph shows 2 separate lines that are made by connecting the points for each city.

In order to use the information given on a graph, you need to understand how to read it. Always read the title of the graph and the labels on the bottom and the left side so that you know what the subject of the graph is. Also, be sure to check out the key, which tells you what the different colors on the graph stand for.

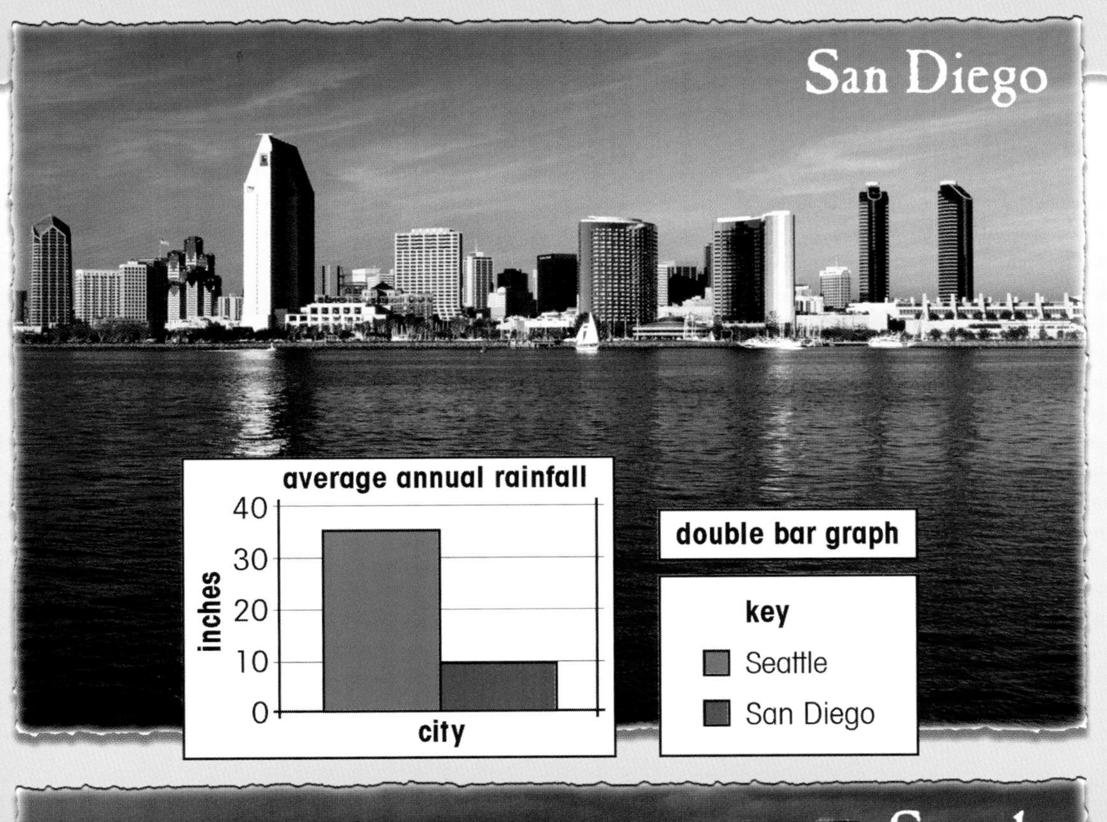

San Diego
average annual rainfall
40
30
20
10
0
inches
city
double bar graph
key
Seattle
San Diego

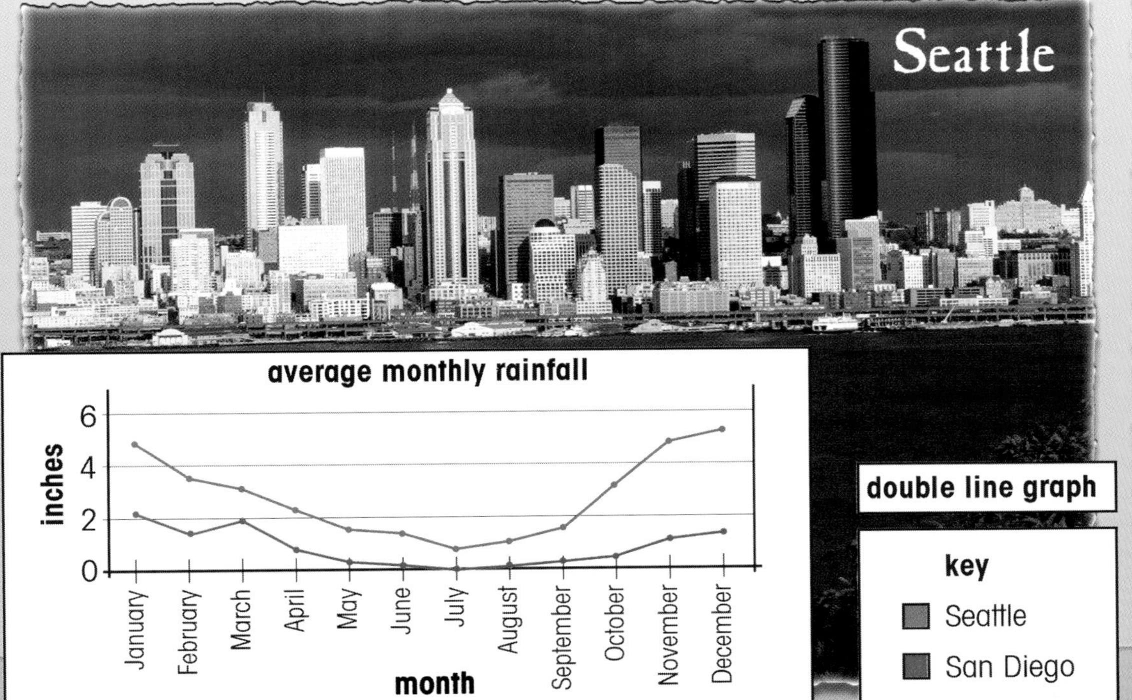

Seattle
average monthly rainfall
6
4
2
0
inches
January
February
March
April
May
June
July
August
September
October
November
December
month
double line graph
key
Seattle
San Diego

A Quick Look at the Amazon and the Sahara

Deep in the jungles of South America, there is a **tropical** ecosystem known as the Amazon rain forest. The Amazon River flows through this region, beginning high in the Andes Mountains of Peru in the western part of South America. It flows all the way through the continent to the Atlantic Ocean on the east side of South America. The Amazon River is about 4,000 miles (6,437 kilometers) long, making it the second longest river in the world. Only the Nile River in Africa is longer.

Around the Amazon River is the world's largest tropical rain forest. A tropical rain forest is a forest that grows in a warm place and receives large amounts of rainfall. The Amazon rain forest takes up $\frac{2}{3}$ of the country of Brazil, covering about 2 million square miles (5.2 million square kilometers).

The Amazon River carries more water than any other river in the world. It has a water flow that is 12 times larger than that of the Mississippi River!

Amazon River
Peru
Amazon
rain forest
Brazil
Andes
Mountains
Pacific
Ocean
Atlantic
Ocean

On another continent, Africa, is the largest desert in the world, the Sahara. A desert is a large area of land that receives little or no rainfall. A desert is often covered with sand, rocks, and dry soil. Because of the lack of rain, deserts do not have a lot of wildlife or trees. The Sahara covers about 3.5 million square miles (9 million square kilometers) and occupies nearly all of northern Africa. Temperatures in the Sahara desert are very hot. The highest temperature ever recorded in the world—136°F (58°C)—was reached in 1922 at a weather station in the Sahara!

Throughout the rest of this book, we'll be comparing and contrasting the Amazon and the Sahara using double bar graphs and double line graphs.

°F = degrees Fahrenheit
°C = degrees Celsius

This double line graph shows that the average monthly rainfall in the Sahara is much lower than the average monthly rainfall in the Amazon. How do you think this affects the regions' ecosystems? We'll find out in the next few chapters.

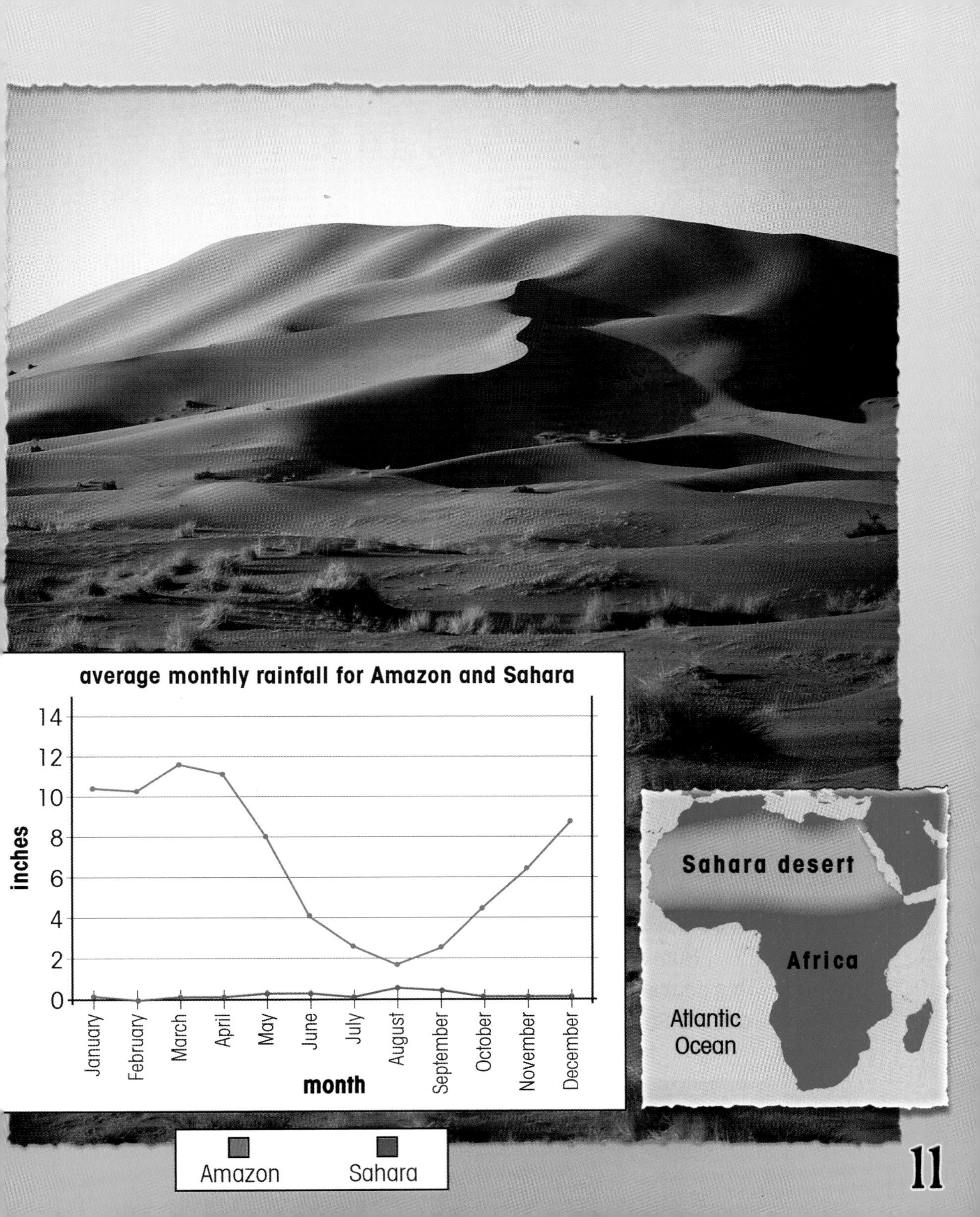
average monthly rainfall for Amazon and Sahara
14
12
10
8
6
4
2
0
inches
January
February
March
April
May
June
July
August
September
October
November
December
month
Amazon
Sahara
Sahara desert
Africa
Atlantic Ocean

A Difference in Climate

The climates of the Sahara desert and the Amazon rain forest are quite different. The Amazon rain forest is hot and **humid** almost all the time. The average temperature of the Amazon is about 80°F (27°C). In the winter, the temperature only drops a few degrees. In the summer, it only rises a few degrees.

The Amazon has 2 seasons: the rainy season and the dry season. The rainy season occurs between December and April, giving the Amazon an average annual rainfall of about 82 inches (2,083 millimeters)! It rains nearly every day in the Amazon, even during the dry season. However, it still rains more during the rainy season. The Amazon is sometimes called "the land of the flooded forest" because of the serious flooding that happens every year from the heavy rainfall.

Humidity is a measure of the "wetness" of the air. This double bar graph compares the humidity in the Amazon and the Sahara. It is much more humid in the Amazon than in the Sahara. Why do you think this is?

average humidity in Amazon and Sahara
100
80
60
40
20
0
%
region
Amazon
Sahara

The Sahara's climate is usually hot and dry. However, winters in the Sahara are colder than winters in the Amazon, and summers are hotter. On a hot summer day, the sand in the Sahara can have a temperature of 170°F (77°C).

Even though the Sahara is dry, there are often thunderstorms in August. The rain can fall so fast during these thunderstorms that sudden, dangerous floods are common. Occasionally, snow falls in the northern areas during cooler months. In the spring, strong, fast winds cause sandstorms throughout most of the region.

Even with the August thunderstorms, the Sahara only receives about 3 to 5 inches (76 to 127 millimeters) of rain a year! Compare this to the annual rainfall in the Amazon. The Amazon gets about 20 times more rain per year than the Sahara. This is why the Amazon is more humid.

This graph shows the average monthly temperatures in both regions. What is the difference in temperature between January in the Amazon and January in the Sahara? According to the graph, it's about 25 degrees.

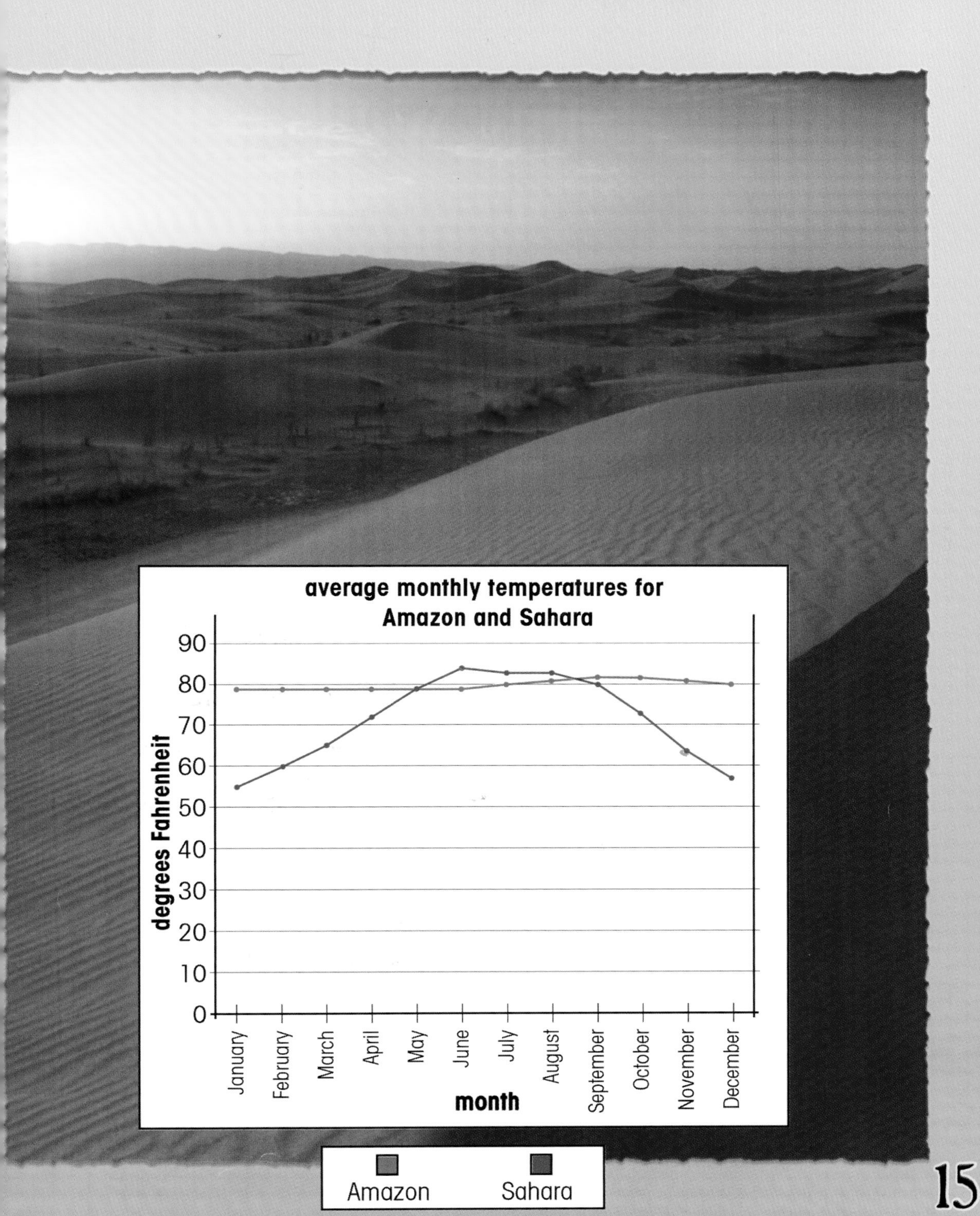
average monthly temperatures for
Amazon and Sahara
degrees Fahrenheit
90
80
70
60
50
40
30
20
10
0
January
February
March
April
May
June
July
August
September
October
November
December
month
Amazon
Sahara

Worldly Wildlife

Although tropical rain forests like the Amazon cover only about 7% of Earth's land surface, they are home to over $\frac{1}{2}$ of the world's wildlife and plant **species**. The Amazon rain forest is home to more types of animals than any other place in the world. Animals of the Amazon include the giant river otter, an **endangered** mammal that is over 6 feet (180 centimeters) long and can weigh more than 70 pounds (32 kilograms). The jaguar, the largest cat in the Amazon, is a silent and fierce hunter. The word "jaguar" means "the beast that kills with 1 bound." The Amazon has hundreds of animals that are found only in its forests, such as the Red and the Black Uakari (wah-kah-REE), the only New World monkeys without a long tail. The Amazon is home to spiders, insects, monkeys, crocodiles, bats, rodents, birds, and hundreds of other living things!

Scientists believe that there are still tens of thousands of species in the Amazon that have yet to be discovered!

jaguar
parrots
giant river otter

The wildlife of the Sahara desert is quite different from the wildlife found in the Amazon. The plants and animals in an ecosystem rely on each other for food, shelter, oxygen, and many other things. The Amazon has much more of most of these things than the Sahara does. The animals that live in the Sahara need to be able to exist on very little water and to endure the desert's heat.

Wildlife in the Sahara includes the addax and the gazelle (two kinds of African **antelope**), the sand cat (the smallest of the wildcats), and the Barbary leopard (an endangered species with only 12 of its kind left in the world). One of the most common desert animals is the camel. A camel can go weeks without food or water! Camels can store up to 80 pounds of fat in the humps on their backs. That fat allows them to survive up to 2 weeks without food.

Which ecosystem has more species in each of these 3 groups of animals? This graph shows that the Amazon has more species of all of the animals listed. This is because most animals need lots of sun, rain, and other plants and animals to eat in order to survive.

camel
gazelle
addax
rain forest vs. desert wildlife
number of species
1,500
1,400
1,300
1,200
1,100
1,000
900
800
700
600
500
400
300
200
100
0
mammals
birds
reptiles
wildlife
Amazon
Sahara

Magical Plants

The Amazon rain forest is packed with tens of thousands of plant species. In the Amazon, you can find the rubber tree. Workers collect the sap from the rubber tree to make rubber the same way people collect sap from maple trees to make maple syrup. Chocolate is made from the cacao bean, which is also found in the Amazon rain forest. Brazil nuts also grow there.

The Amazon is also thought of as the world's largest natural **pharmacy**! Deep in the Amazon rain forest, hundreds of natural medicines have been discovered. Quinine (KWY-nine) comes from the bark of the cinchona (sihn-KOH-nuh) tree and is used to treat people who have an illness called **malaria**. Cat's Claw is a plant that can help the body fight off diseases. These are just 2 of the hundreds of plants in the Amazon that can be used to make medicines.

The Amazon rain forest is home to more trees and plants than anywhere else in the world. Scientists believe that there are still tens of thousands of species left undiscovered. Some people think the cure for cancer might be discovered in the Amazon one day!

cacao beans
rubber
tree
cinchona tree

The climate of the Sahara desert does not allow much plant growth. However, the species that do live there are interesting because of their ability to **adapt** to sudden changes in the Sahara's climate. It is estimated that in all of the 3.5 million square miles (9 million square kilometers) that make up the Sahara, only 500 plant species exist!

Grasses, shrubs, cactuses, palm trees, and wildflowers make up the majority of the Sahara's plant life. Some of these plants have roots that reach down 80 feet (24 meters) in order to soak up moisture from deep below the surface! Many plants in the Sahara depend on rain to live. However, rainfall does not provide a constant supply of water for the plants. Most Saharan plants grow when the rain comes, but once the dry season arrives, they dry up and die. Their seeds, which are spread by the wind and animals while the plants are in bloom, begin to grow when the rainy season comes again.

This bar graph compares the number of plant species that can be found in the Amazon and in the Sahara. Why do you think there are so many more plant species in the Amazon than in the Sahara?

rain forest vs. desert plant life
number of species
80,000
75,000
70,000
65,000
60,000
55,000
50,000
45,000
40,000
35,000
30,000
25,000
20,000
15,000
10,000
5,000
0
region
Amazon
Sahara

The People of the Amazon and the Sahara

Although Europeans discovered the treasures of the Amazon only a few centuries ago, its native people have enjoyed its wonders for about 12,000 years. The native people of the past knew the animals of the jungle by their footprints and their smells. They knew which plants cured which sicknesses, which fruits were safe to eat, and which were deadly. Today, the people of the Amazon still use hundreds of different kinds of plants to make canoes, houses, hammocks, baskets, and other things they need to survive.

During the 1500s, Europeans discovered Brazil and the mighty Amazon River. Since then, hundreds of the native tribes have disappeared or died out. When gold miners invaded the Yanomami (yah-nuh-MAH-mee) land in the 1980s, they caused great harm to the tribe's population. The outsiders carried with them diseases like malaria, which were unknown in the Amazon. About 2,000 Yanomami people died within 10 years of the miners' arrival.

Yanomami tribe members

Today, there are about 10,000 Yanomami people left in the Amazon rain forest. The Brazilian government has passed laws to keep people from destroying Yanomami land, but gold miners and logging companies continue to do so.

About 2 million people call the Sahara desert their home. Around $\frac{2}{3}$ of these people live in the areas of the desert called oases (oh-AY-seez). An oasis is a place in the desert where the underground water rises and makes it possible for things to grow. Some people who live in oases even raise crops such as dates and wheat.

The remaining $\frac{1}{3}$ of the people who live in the Sahara are nomads. Nomads are people who spend their lives traveling. Unlike the Amazon, the Sahara does not have enough food and water for survival, so these nomadic people travel from place to place for these things. They wear long robes and **turbans** to protect their skin from the strong Saharan sun. The nomadic people of the Sahara are herders and traders. Bedouins (BEH-duh-wenz) and Tuareg (TWAH-reg) are just 2 of the groups of nomads that still live in the Sahara.

As the graph shows, many more people live in the Amazon than in the Sahara. Why do you think this is?

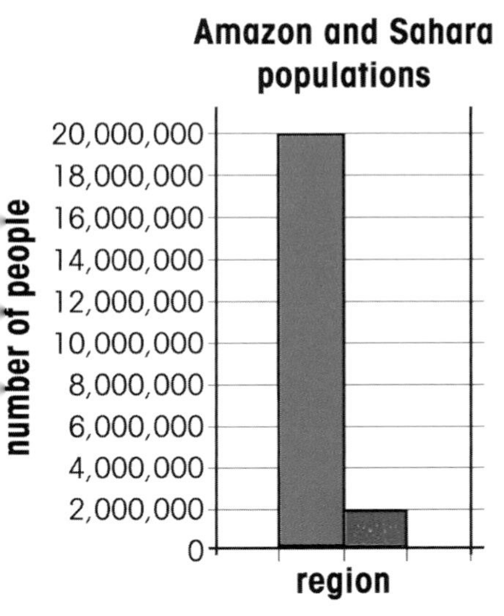

Amazon and Sahara populations
20,000,000
18,000,000
16,000,000
14,000,000
12,000,000
10,000,000
8,000,000
6,000,000
4,000,000
2,000,000
0
number of people
region

Amazon

Sahara

Changing Ecosystems

The Sahara desert and the Amazon rain forest have experienced many changes throughout history. As a part of the world's community of ecosystems, over the years these 2 regions have reacted to changes in the world around them. Differences in the weather, atmosphere, and living things all around the world affect each and every ecosystem.

The Sahara desert has experienced changes in its wildlife, plant life, climate, and area for thousands of years. Over the past 20 years, the Sahara has shrunk and then grown again. In the 1980s, a lack of rain caused the southern border of the Sahara to spread farther south. Today, however, the southern border is turning green with plant life again. Scientists think this is because the rainfall in that area has increased and because better farming methods are used today.

In 2002, satellite photographs showed that a 4,000 mile (6,437 kilometer) stretch of the Sahara desert—from the Atlantic Ocean to the Red Sea—is turning green again.

Since the late 1970s, the size of the Amazon rain forest has decreased because of mining, logging, and the building of modern roads. Although the Amazon rain forest has been **deforested** over the years, some people have realized the importance of the Amazon ecosystem and have been trying to educate others about how to help protect the Amazon from more damage. The rate of deforestation has declined in recent years, but we all need to work together to keep the Amazon rain forest alive.

If mathematicians had not invented ways for us to study data, we would not be able to measure the dangers facing these ecosystems. Knowing how to read graphs can help us understand so much about the environment and the world around us. It can even help us save a rain forest!

Glossary

adapt (uh-DAPT) To change in order to fit in or live.

antelope (AN-tuh-lohp) An animal of Africa and Asia that is similar to a deer.

data (DAY-tuh) Information, usually facts or numbers, used for drawing conclusions.

deforest (dee-FOR-uhst) To clear the trees from an area of land.

ecosystem (EE-koh-sihs-tuhm) A community of living plants and animals.

endangered (in-DAYN-juhrd) At risk of not living anymore.

humid (HYOO-mid) Damp and moist.

malaria (muh-LAIR-ee-uh) A disease that causes chills, fever, and sweating. Malaria is carried from person to person by insects called mosquitoes.

mathematician (math-muh-TIH-shun) A person who is an expert at math.

pharmacy (FAR-muh-see) A place where you can get medicine for sicknesses.

species (SPEE-sheez) A group of plants or animals with similar features.

tropical (TRAH-puh-kuhl) Warm year-round.

turban (TUR-buhn) A headdress made of a long piece of cloth that is wound around the head.

unit (YOU-nuht) A standard quantity used for measurement.

Index